ENVIRO-CHALLENGE

teri

The Energy and Resources Institute

An imprint of The Energy and Resources Institute

First reprint 2017
Second reprint 2018

First published in 2012 by
The Energy and Resources Institute
TERI Press
Darbari Seth Block, IHC Complex, Lodhi Road, New Delhi 110 003, India
Tel. 2468 2100/4150 4900, Fax: 2468 2144/2468 2145
India +91 ■ Delhi (0)11
Email: teripress@teri.res.in ■ Website: http://bookstore.teriin.org

ISBN 978-81-7993-458-6

Publishing Head: Anupama Jauhry
Content Development: Educating Youth for Sustainable Development Division, TERI
Editorial Team: Rupak Ghosh
Design and Layout Teams: Santosh Gautam, Vijay Kumar, Rajiv Sharma
Production Team: R K Joshi, R Ajith Kumar, Aman Sachdeva

Printed and bound in India

This book is printed on recycled paper.

A note to the readers

The purpose behind this compilation of over 500 quiz questions is to improve and expand the understanding of issues pertaining to environment education, primarily at the school level. This is because, only when the understanding of young readers is steered towards pressing environment-related matters, can they be guided towards a path of 'joyful learning'—the precise idea of this book.

'Enviro-challenge' is also an endeavour to supply schools a ready resource kit on environmental topics, with particular emphasis on mainstreaming environment in the formal education structure. While assessing the 'environment quotient' of students in the 12-15 age group, it aims to raise awareness on a wide range of areas, including water, energy, air, biodiversity, global warming, and sustainable development, among others.

As the readers turn the pages, it is hoped they will feel encouraged to start acting as catalyst and agents of change, further sensitizing their peers, siblings, parents, and neighbours on these pivotal issues.

- Educating Youth for Sustainable Development (EYSD) Division, TERI

Contents

Section I
Multiple-Choice Questions

1. Across Africa, Asia, and Latin America, this plant is a valuable source of food, furniture, and construction materials. In China, it is known as 'Friend of the people'. Name the plant.
   a) Bamboo
   b) Teak
   c) Mahogany
   d) Jute

2. Biogas is produced when anaerobic bacteria digest organic matter in the absence of oxygen. What is this process called?
   a) Anaerobic reduction
   b) Anaerobic digestion
   c) Anaerobic oxidation
   d) Anaerobic drying

3. The spraying of which pesticide in the cashew plantations at Kasargod district of Kerala has resulted in debilitating diseases like mental retardation and cerebral palsy?
   a) DDT
   b) Endosulphan
   c) Malathion
   d) Pyrethrin

4. Some greenhouse gases (GHGs) occur naturally in the atmosphere, while others result from human activities. Naturally occurring GHGs include water vapour, CO_2, methane, nitrous oxide, and ozone. Of the following, which gas is more effective than CO_2 in trapping heat in the atmosphere?
   a) Methane
   b) Sulphur dioxide
   c) Ozone
   d) Water vapour

5. It is a means of acquiring information applying airborne equipment and techniques to determine the characteristics of an area. Aerial photographs from aircraft and satellite are the most common forms used. It usually relies upon measurement of electromagnetic energy reflected or emitted from the features of interest. What are we talking about?
   a) Remote sensing
   b) Electromagnetic sensing
   c) Photography
   d) Topography

6. In the ceramics industry, certain mineral elements and compounds are highly toxic. They are emitted by raw materials used in various steps of the firing process. Most of the ceramic colouring and waste are hazardous to the environment due to their high metal content. Which of the following are some of the toxins that are released?
   a) Chlorine
   b) Fluoride
   c) Both chlorine and fluoride
   d) None of the above

7. Forest fires can be caused naturally (for instance, by lightning) or by man. The burning of trees in a forest increases which of the following gases in the atmosphere?
   a) Ozone
   b) Carbon dioxide
   c) Sulphur dioxide
   d) Methane

8. An alarming fact is that household detergents are nearly always used in far greater amounts than necessary. Which particular pollutants are present in household detergents?
   a) Phosphates and polyphosphates
   b) Sulphates
   c) Nitrates
   d) Carbon dioxide

9. Darwin's theory of evolution challenged the Genesis's account of special creation of species. It implied a continuity of humans and animals, of within-species and between-species variation, not only in body but also in behaviour. According to Darwin's theory of evolution, differences between species may be the result of which of the following?
   a) The disuse of body structures
   b) The transmission of acquired characteristics
   c) Natural selection
   d) Mutagenic agents

10. Most flowering plants depend on insects like honeybees and butterflies to transport certain 'grains' from one part of a flower to another. What are these 'grains' better known as?
    a) Stomata
    b) Pollen
    c) Lenticels
    d) Columnar epithelial tissue

11. Cytokinin, the plant growth hormone, is responsible for which process in plants?
    a) Cell division
    b) Photosynthesis
    c) Budding
    d) Reproduction

12. Ants are sometimes involved in a symbiotic relationship with other insects. These insects are considered to be major crop pests, sucking the fluid content from a plant's phloem tissues. The ants provide these insects with protection from other insects; and the insects, in turn, act as a source of food for the ants. Which insects are we referring to?
    a) Cockroaches
    b) Aphids
    c) Butterflies
    d) Ladybugs (ladybirds)

13. Tigers are an endangered species, and are being protected all around the world. In which year was 'Project Tiger' initiated?
    a) 1975
    b) 1973
    c) 1977
    d) 1971

14. A crocodile's stomach is divided into two chambers. The first one is powerful and muscular. This is where stones or gastroliths are found in most adult salt-water crocodiles. The other chamber has the most acidic digestive system of any animal. It can digest mostly everything—from bones and feathers to horns. Why do most adult salt-water crocodiles have stones in their stomach?
    a) They help a crocodile balance in water
    b) Stones help crush food in the body
    c) Crocodiles swallow them by mistake
    d) They help in the digestion of food

15. An insecticide is a pesticide used against insects in all stages of development. The appearance of new insect varieties that are resistant to insecticides is an illustration of...?
    a) Thigmotropism
    b) Inheritance of acquired traits
    c) Ecological succession
    d) Natural selection

16. Mammals are a diverse group inhabiting a wide variety of habitats. Their characteristics are the result of a range of adaptations that allow them to survive in the numerous environmental conditions they live in. Which of the following represents a mammal that is adapted to fly?
   a) Penguin
   b) Eagle
   c) Bat
   d) Bee

17. This plant, native to Sri Lanka, is considered an important spice. It has been in vogue since the days of the ancient Egyptian civilization. It was also used in embalming, where body cavities were filled with spiced preservatives. Derived from strips of bark rolled one in another, its best varieties are pale and parchment-like in appearance. Which spice are we talking about?
   a) Cinnamon
   b) Sandalwood
   c) Eucalyptus
   d) Tea

18. A strong-smelling secretion produced by the glands of this animal has been used in perfumes and traditional medicine in Asia for over 5,000 years. Protecting this species is therefore important from both commercial and conservation perspectives. Name this animal.
   a) Chiru
   b) Musk deer
   c) Reindeer
   d) Ibex

19. Plants need nitrogen to grow, but atmospheric nitrogen cannot be used by plants. What is the most common process by which nitrogen is made available to plants?
   a) Breakdown of water
   b) Bacterial action
   c) Photosynthesis
   d) Lightning

20. Plants always tend to adapt to the climatic and environmental conditions of the area they are found in to protect themselves or for their growth. What is the typical texture of the bark of a tree in the rain forest?
   a) Thick: to protect against heat and water loss
   b) Thin and smooth: to protect against creeping plants
   c) It varies: making it easy to tell one tree from another
   d) Thick and smooth: to ensure creeping plants grow on it

21. This animal is regarded as precious and majestic since the time of *Samudra Manthan*. Later, it became the divine vehicle of Indra. Which animal are we talking about?
    a) Lion
    b) Deer
    c) Mouse
    d) Elephant

22. This is the largest bird park in the Asia-Pacific region, with more than 9,000 birds across 600 species. The park, covering an area of 50 acres, claims to contain the world's largest and most impressive collection of birds. Which park is this?
    a) Lambir Hills National Park (Sarawak, East Malaysia)
    b) Jurong Bird Park (Singapore)
    c) Khao Sok National Park (southern Thailand)
    d) Kuala Gula Bird Sanctuary (Malaysia)

23. India's first national park, the Hailey National Park, named after Governor Malcolm Hailey, was established in August 1936 under the UP National Park Act. Situated in the foothills of the Himalayas and famous for its population of tigers, what is this park currently known as?
    a) Malcolm Hailey National Park
    b) Jim Corbett National Park
    c) Sariska National Park
    d) Bandhavgarh National Park

24. Name the bird, which derives its name from its habit of perching on mammals, and picks up ticks, pests, larvae, and pesticides. It shares a symbiotic relationship with its buddy animals and feeds while sitting on their back.
    a) Pigeon
    b) Oxpecker
    c) Sparrow
    d) Crow

25. What is the symbiotic relationship between the crocodile and the Egyptian Plover bird known as, wherein the bird is allowed to sit within the jaw of the crocodile?
    a) Parasitism
    b) Mutualism
    c) Commensalism
    d) Amensalism

26. What is the popular term for salt-tolerant forest ecosystems found mainly in tropical and sub-tropical inter-tidal regions of the world?
    a) Mangroves
    b) Wetlands
    c) Marshlands
    d) All of the above

27. Which crocodile bank was founded in India in 1976 by the famous herpetologist Romulus Whitaker?
    a) Kolkata Crocodile Bank
    b) Chennai Crocodile Bank
    c) Madras Crocodile Bank
    d) The Crocodile Bank

28. This sirenian mammal was first described by the naturalist Georg Steller during an expedition led by explorer Vitus Bering. Within 27 years of its discovery, this once-abundant species in the North Pacific was hunted to extinction. Name this mammal.
    a) Sea horse
    b) Shark
    c) Whale
    d) Sea cow

29. The major threats to biodiversity worldwide can be categorized using the HIPPO dilemma. What does HIPPO stand for?
    a) Hippopotamuses
    b) Habitat Loss, Introduced Species, Pollution, Population, Over-consumption
    c) Hunting, Isolation, People, Propaganda, Opulence
    d) Harmful ultraviolet rays, International trade, Politics, Power production, Oxygen deprivation

30. What is the interaction between two species in which one is benefited but the other is neither helped nor benefited (for example, creeper plants on trees) called?
    a) Mutualism
    b) Commensalism
    c) Parasitism
    d) Symbiosis

31. Gahirmatha rookery, located along the northern Orissa coast, hosts the largest-known nesting concentration of which of the following species?
    a) Loggerhead turtle
    b) Olive Ridley turtle
    c) Great Indian Bustard
    d) Mugger crocodile

32. Wind turbines can be divided into two categories based on the axis about which the turbine rotates. Turbines that rotate about a horizontal axis are most common. Vertical-axis turbines are less frequently used. Why do wind turbines using aerodynamic lift produce more energy for a given swept area than wind turbines using aerodynamic drag?
    a) The lifting force pushes the blade in the direction the wind is already blowing
    b) The lifting force is roughly perpendicular to the local flow field
    c) Lift produces more torque
    d) Drag devices capture more energy because of more friction with blade surfaces

33. Coal liquefaction is one of the technologies that will limit escalation of oil prices and mitigate the alleged effects of peak oil, provided the additional CO_2 released in the process does not catastrophically accelerate global warming or perpetuate adverse climate effects. Coal liquefaction refers to the process of...
    a) Washing coal to remove impurities
    b) Turning coal into syncrude
    c) Turning coal into natural gas
    d) None of the above

34. A passive solar heating system collects energy from the sun and uses it to heat a space directly, such as a sunroom, or to heat a fluid. What do we call the heat storage material used in passive solar heating systems, which radiates heat to the surroundings after the sun goes down?
    a) Thermal mass
    b) Collector
    c) Absorber
    d) None of the above

35. On 27 May 1927, for the first time in the history of Sikkim, a micro-hydel station was commissioned, with an installed capacity of 50 kW, which was distributed through a 3.3 kV overhead transmission line. This was essentially for the members of the royal family and the town of Gangtok. On which river was the station situated?
    a) Ranikhola
    b) Ranikhet
    c) Sichey Busty
    d) Teesta

36. A nuclear reactor produces and controls the release of energy from splitting atoms of certain elements. In a nuclear power reactor, the energy released is used as heat to make steam to generate electricity. The energy released from continuous fission of the atoms of the fuel is harnessed as heat in either a gas or water, and is used to produce steam. The steam is used to drive the turbines, which produce electricity. There are several types of nuclear reactors in use today. Which one of these reactors makes more fissile fuel than it consumes?
    a) Boiling water reactor
    b) Pressurized water reactor
    c) Advanced gas-cooled reactor
    d) Breeder reactor

37. FutureGen is an initiative to build and operate the world's first coal-based power plant that removes and sequesters CO_2 while producing electricity and hydrogen. Which was the first country to join the US on the government steering committee for the FutureGen Initiative?
    a) India
    b) Canada
    c) United Kingdom
    d) Japan

38. Geothermal power is the use of geothermal heat for electricity generation. It is often referred to as a form of renewable energy, but because the heat at any location can eventually be depleted, it technically may not be strictly renewable. Geothermal-generated electricity was first produced at Larderello, in 1904. In which country is Larderello located?
    a) Spain
    b) Portugal
    c) Italy
    d) Greece

39. A normal light bulb is also known as an incandescent bulb. It contains a very thin tungsten filament. On the other hand, a fluorescent bulb has electrodes at both ends of a tube containing argon and mercury vapour. When it comes to converting electricity into light, how do compact fluorescent bulbs compare to incandescent light bulbs?
    a) Compact fluorescent bulbs are more efficient
    b) Compact fluorescent bulbs have the same efficiency as incandescent bulbs
    c) Compact fluorescent bulbs are less efficient
    d) None of the above

40. The direction the glass faces is the most important factor in keeping a room cool during summer. If a room in your house had glass windows that faced the east and west instead of south, how much heat would the room get during the summer?
    a) The same amount of heat
    b) Less heat
    c) More heat
    d) None of the above

41. A ceiling fan uses about 98 per cent less energy than most central air-conditioners. As compared to a light bulb, a bulb of how many watts will consume the same amount of energy?
    a) 40-watt bulb
    b) 60-watt bulb
    c) 20-watt bulb
    d) 100-watt bulb

42. People have a habit of keeping a lot of their appliances on the standby mode throughout the day. In a household, at the end of the day, which of the following appliances left behind on standby mode will consume more electricity?
    a) TV set
    b) Toaster
    c) Refrigerator
    d) Microwave oven

43. The first commercial nuclear power stations started operating in the 1950s. There are now 435 commercial nuclear power reactors operating in 30 countries, with over 368,000 MWe of total capacity. Which country produces the maximum commercial nuclear power?
    a) United Kingdom
    b) Canada
    c) Japan
    d) United States of America

44. Saudi Arabia has the world's largest known reserves of this resource. It is a non- renewable source of energy that is used for transportation purposes. It is refined to make gasoline. Can you identify it?
    a) Natural gas
    b) Petroleum
    c) Hydropower
    d) Coal

45. Which colourless, odourless fossil fuel, with chemical name methane, is considered to be the cleanest burning fuel?
    a) Natural gas
    b) Solar energy
    c) Biomass
    d) Geothermal

46. This portable source of heat energy is normally stored under pressure in tanks and bottles. It is also buried underground. Which source of energy are we referring to?
    a) Propane
    b) Biomass
    c) Petroleum
    d) Solar energy

47. The December 2004 earthquake and the subsequent tsunami were caused by the collision of two continental plates off the western coast of Sumatra. Can you name the plates?
    a) Pacific and Australian
    b) African and Eurasian
    c) Eurasian and Caribbean
    d) Indian and Burmese

48. Pliny the Younger witnessed the eruption of Mount Vesuvius from 20 km away and recorded the event. This was the earliest scientific account of a volcanic eruption. In approximately which year did this volcanic eruption take place?
    a) AD 1247
    b) AD 1602
    c) 2000 BC
    d) AD 79

49. After the eruption of Krakatoa in 1883 in Indonesia, more than 36,000 people were killed. What caused these deaths?
    a) Tornado
    b) Lava
    c) Typhoon
    d) Tsunami

50. On 30 June 1908, a natural disaster occurred in the Stony Tunguska River Valley of Siberia. It flattened trees and destroyed over 6,000 square kilometres of forests. What type of natural disaster was it?
   a) A flood
   b) An earthquake
   c) A comet
   d) A forest fire

51. Biodiesel refers to diesel-equivalent biofuel or vegidiesel that is generally produced from vegetable oil or animal fat. Even though several different kinds of fuels are considered to be biodiesel, commonly biodiesel refers to which of these?
   a) Ester
   b) Oxygenate (produced from oil and methanol)
   c) Alkane (non-oxygenate) biodiesel, i.e. BTL (biomass-to-liquid) fuel
   d) All of the above

52. The Honda Insight and the Toyota Prius are car models that represent a breakthrough in environmental technology, and are referred to as 'hybrid cars'. They are cleaner and more fuel-efficient forms of transportation. What are these cars powered by?
   a) Petrol and diesel
   b) Electricity
   c) Petrol and electricity
   d) Biofuel

53. In 1844, the following appeared in a Brooklyn newspaper: '...the iron horse with its lungs of brass and sinews of steel...puffing volumes of smoke and flame from its nostrils...' What did the nickname 'iron horse' refer to?
   a) Trains
   b) Automobiles
   c) Steamships
   d) Aircrafts

54. Which one among the following vehicle categories is responsible for the highest pollution emissions?
   a) Cars
   b) SUVs
   c) Motorbikes
   d) Scooters

55. This gas-charged igneous rock is basically lava froth. It looks solid, but is full of pores and spaces, and weighs very little. It is easily crushed, and the bubbles in it are small and irregular. Which rock are we referring to?
    a) Pumice
    b) Granite
    c) Basalt
    d) Quartz

56. Rock and soil are altered physically by disintegrating, and chemically by decomposing. Freeze, thaw, frost action, root pry, and abrasion are all types of...
    a) Chemical weathering
    b) Physical weathering
    c) Deposition
    d) Erosion

57. Mother Nature makes glass each time lightning strikes the Earth's surface, provided the soil composition is suitable for making glass (contains quartz). Hollow, glass tubes formed in soil by lightning strikes is known as...
    a) Fragrite
    b) Obsidian
    c) Fulgurite
    d) Onyx

58. On 2 December 1942, at the University of Chicago, Enrico Fermi succeeded in producing a chain reaction in a pile of uranium. This marked the beginning of which age?
    a) The nuclear age
    b) The nuclear fission age
    c) The nuclear fusion age
    d) None of the above

59. The Birds of America is a book by a famed naturalist and painter, and is considered to be the first birding guide. The book contains over 400 illustrations of birds of America drawn by the author during 13 years of travel. It also carries images of six birds that are now extinct: the Great Auk, Carolina Parakeet, Passenger Pigeon, Labrador Duck, Pinnated Grouse, and the Esquimaux Curlew. Who is the author of The Birds of America?
    a) John James Audubon
    b) Rachel Carson
    c) Robert Peck
    d) Jack Nicholson

60. Can you name this biennial report series published by the United Nations Environment Programme (UNEP) that stipulates the world environmental priorities, identifying the major issues and possible solutions through policy directives?
    a) GEO (Global Environment Outlook)
    b) SoE (State of Environment)
    c) EPR (Environmental Priorities Report)
    d) None of the above

61. In 2000, ministers of environment and other delegates met for the First Global Ministerial Environment Forum. It was a follow-up to the UN General Assembly resolution of 'enabling the world's environment ministers to review important and emerging environmental issues and to chart the course for the future'. What is the first forum on international environmental governance referred to as?
    a) 2000 - Malmo Declaration
    b) UNICLOS Convention
    c) Convention on Biodiversity
    d) Delhi Declaration

62. The United Nations has earmarked the decade 2005–2014 towards which of the following themes?
    a) Decade of the World's Indigenous People
    b) Decade of Education for Sustainable Development
    c) Decade of Recovery and Sustainable Development of the Affected Regions
    d) Decade for Deserts and the Fight against Desertification

63. On which of the following days was the World Migratory Bird Day observed in 2011?
    a) 14–15 April
    b) 14–15 June
    c) 14–15 May
    d) 14–15 December

64. The Ramsar Convention held at Ramsar, Iran in 1971 aims at protection of which of the following?
    a) Wetlands
    b) Biodiversity
    c) Oceans
    d) Forests

65. Which of the following is an international umbrella organization established in 1972 for organic farming bodies?
    a) FAO
    b) IFOAM
    c) IFAD
    d) WFP

66. Which country was chosen as the host for the United Nations Conference on Sustainable Development (UNCSD) in 2012?
    a) Sweden
    b) India
    c) South Africa
    d) Brazil

67. Oxygen-demanding wastes serve as food supplies for decomposers. Consumption of dissolved oxygen in water by decomposers can render a stream uninhabitable for various types of aquatic flora and fauna. Oxygen-demanding wastes are contained in effluent from which of the following?
    a) Agricultural runoff
    b) Power plants
    c) Inorganic industrial pollutants
    d) Mining operations

68. When clearing up leaves and other garbage, it is advisable to segregate and compost the leaves and vegetable waste, rather than burning leaf piles. What does leaf smoke contain?
    a) Particulate matter
    b) Hydrocarbons
    c) Carbon monoxide
    d) All of the above

69. Composting is a good practice for managing waste. What are the potential benefits of composting?
    a) Helps reduce soil compaction and erosion
    b) Helps soil retain moisture and nutrients, while increasing soil fertility
    c) Can also serve as organic fertilizer
    d) All of the above

70. Which of the following is produced on burning plastic waste that is carcinogenic by nature?
    a) Lead
    b) Dioxins
    c) Mercury
    d) Asbestos

71. In 2005, the decommissioned French warship Clemenceau was headed for the Alang ship breaking yard on the Gujarat coast with tonnes of toxic waste. Which waste was it carrying?
   a) Lead
   b) Dioxins
   c) Mercury
   d) Asbestos

72. The study of the ocean, comprising all knowledge with regard to its physical boundaries, the chemistry and physics of seawater, and marine biology is known as oceanography. Who is the 'father' of oceanography?
   a) James Cook
   b) Matthew Maury
   c) Jacques Cousteau
   d) Harry Hess

73. The Stockholm Water Prize is an international award founded in the 1990s, and is presented annually to an individual, organization, or institute for outstanding water-related activities. Name the Indian-born Canadian who won this prestigious prize in 2006 for challenging the global status quo on water.
   a) Asit K Biswas
   b) Dr Shiv Chopra
   c) Dr Ronald Micetich
   d) Lisa Ray

74. It was one of the most important and significant factors that led to the prosperity and success of the world's early civilizations. Which factor are we referring to?
   a) Formation of governments
   b) Development of monotheism
   c) Location on river valleys
   d) Building of planned cities

75. Water cycle is the process of natural circulation of water in solid, liquid, or gaseous form in the biosphere. The continuous process of evapo-transpiration, condensation, and precipitation makes up the water cycle. What is the process by which plants give off water vapour to the atmosphere known as?
   a) Evaporation
   b) Transpiration
   c) Condensation
   d) Precipitation

76. In water purification, impurities are caused to coagulate and are removed as they settle at the bottom of the container. Which form of aluminium is used in waste water treatment plants as a coagulant to remove particles?
    a) Aluminium phosphate
    b) Aluminium sulphate
    c) Aluminium oxide
    d) Aluminium nitrate

77. All of the following are very long rivers that flow in different parts of the world. Of these, whose drainage basin is located entirely in one country?
    a) Amazon
    b) Yangtze
    c) Ganga
    d) Nile

78. A drainage basin (also known as a watershed) is a region of land where water from rain or snowmelt drains downhill into a body of water, such as a river, lake, estuary, wetland, sea, or ocean. Name the largest river drainage basin in Europe.
    a) Volga
    b) Danube
    c) Rhine
    d) Thames

79. This element has led to widespread contamination in the groundwater of West Bengal and Bangladesh, and has been detected to be present at much higher levels than WHO's permissible limit of 0.01 mg/litre. Its long-term exposure via drinking water causes cancer of the skin, lungs, urinary bladder, and kidneys; as well as skin pigmentation and thickening (hyperkeratosis). What is the name of this element?
    a) Lead
    b) Arsenic
    c) Fluoride
    d) Pesticides

80. Most lakes and streams have a pH level between 6 and 8. How does acid rain increase the acidity of water environments?
    a) By raising the pH level
    b) By lowering the pH level
    c) By keeping the pH level same
    d) None of the above

81. This ship canal project across the Palk Strait between India and Sri Lanka was inaugurated some time ago. Though the government has assured that dredging will be done 20 kilometres away from the ecologically sensitive zone, environmentalists and fishermen are crying foul as they claim the churning of the ocean floor will destroy the bio-reserve, and with it, their livelihood. Name this canal project.
   a) Sethusamudram
   b) Ramsetu
   c) Sethulanka
   d) Sethusagar

82. Many earth scientists agree with the hypothesis that both the atmosphere and the oceans have accumulated gradually via a process of 'degassing' of the Earth's interior. Which of the following gases found in ocean water do not come from the atmosphere?
   a) Carbon dioxide
   b) Chlorine
   c) Nitrogen
   d) Oxygen

83. Warm surface currents invariably flow from the Tropics to the higher latitudes, driven mainly by atmospheric winds, as well as the Earth's rotation. These currents affect the water temperature along which of the following coast of continents?
   a) North
   b) West
   c) South
   d) East

84. This oil spill is the largest oil spill in the history of the US and the petroleum industry. It began on 20 April 2010 and stopped on 15 July 2010 after the well head was capped. Which oil spill are we referring to?
   a) San Francisco oil spill
   b) US Navy oil spill
   c) Deepwater horizon oil spill / BP oil spill / Macondo blowout
   d) None of the above

85. The Deepwater Horizon spill caused extensive damage to the marine habitat. It is estimated that approximately 5 million barrels of oil spilled, which amounts to about 6 per cent of the daily global oil consumption. Where did this marine tragedy happen?
   a) San Francisco Bay
   b) Gulf of Mexico
   c) Off Mumbai coast
   d) Prince William Sound, Alaska

86. The famous Exxon Valdez case is linked to which of the following events?
    a) Oil spill
    b) Chemical spill
    c) Nuclear disaster
    d) Forest fire

87. A desktop computer contains various chemicals. Which of the following chemicals is NOT contained in an ordinary computer and monitor?
    a) Mercury
    b) Dioxins
    c) Arsenic
    d) Brominated flame retardants

88. This compound, a mixture of up to 209 congeners, is used as a coolant and lubricant in transformers, capacitors, and other electrical equipment. Once released in the environment, it does not readily break down and keeps circulating in the atmosphere for many years. It has been linked to cancer and is known to accumulate in the fatty tissues of fish, animals, and people. Name the compound.
    a) Pentachloro benzene
    b) Polychlorinated biphenyls
    c) Pendimethalin
    d) None of the above

89. A popular jewellery component is formed when an irritant works its way into a particular species of mollusc. As a defence mechanism, the mollusc secretes a fluid to coat the irritant. Several layers of this coating (known as nacre) are deposited on the irritant until a lustrous substance is formed. Name the substance.
    a) Natural pearl
    b) Shell
    c) Diamond
    d) Ruby

90. Most conventional mosquito repellents contain a known persistent organic pollutant, which, though an effective repellent, is also an eye irritant. It is also known to cause blisters and rashes on some users. It is now also being associated with lethargy, confusion, disorientation, and mood swings. This pollutant stays in the soil for years and is found in waterways. What is the scientific name of this pollutant?
    a) Hexachlorobenzene
    b) Furans
    c) N, N- Diethyl-3- Methylbenzamide (DEET)
    d) Fluoride

91. Plastic toys made of PVC (polyvinyl chlorides) contain dangerous toxic additives, which could leak out of the toy and be ingested by children. Phthalates, the plasticizers added to PVC to make it softer and more elastic, when leached out of the toy could pose potential health risk. Described as the 'most abundant manmade environmental pollutant', the phthalate most commonly found in toys, is a known toxic hazard. Name the phthalate found in toys.
    a) Diisononyl phthalate
    b) Diisodecyl phthalate
    c) Both a and b
    d) Dimethyl phthalate

92. Two cement plants located directly above India's longest limestone caves, near the Norpuh Reserve Forests, have allegedly come up without the necessary environmental and other clearances from relevant departments. As a result, the caves are under threat. The mining will also impact critical wildlife habitats. Which state in north-east India is home to these caves and also a hub of cement companies?
    a) Assam
    b) Meghalaya
    c) Mizoram
    d) Nagaland

93. One of the following could be used at sites where oil spills have occurred and the soil has been damaged. The process encourages the growth of bacteria that break down petroleum products in the soil. What are we referring to?
    a) Hydrogen releasing compound
    b) Oxygen releasing compound
    c) Nitrogen supplementation
    d) H_2O saturation

94. The first satellite primarily built for ocean applications by ISRO's PSLV (Polar Satellite Launch Vehicle) photographed the tsunami waves crashing over Car Nicobar island between 7:30 am and 7:50 am. Name the satellite.
    a) Rohini (RS-1)
    b) IRS-P4
    c) Aryabhatta
    d) Bhaskara

95. The properties and characteristics of a particular member of the plastics family cause carbonated drinks to bubble more in plastic cups. Its ability to contain carbon dioxide makes it the material of choice for carbonated beverage containers. Can you name it?
   a) PET (polyethylene terephthalate)
   b) LDPE (low-density polyethylene)
   c) PP (polypropylene)
   d) HDPE (high-density polyethylene)

96. Twice a year, the Earth experiences solstices—once, when the sun is directly over the Tropic of Cancer, and then again when the sun is over the Tropic of Capricorn. On 21 June, certain locations have 24 hours of daylight. These locations are all between which of the following latitudes?
   a) 0° and 23½°N
   b) 66½°N and 90°N
   c) 23½°N and 47°N
   d) 47°N and 66½°N

97. The human development index, used as a general indicator of human well-being, is based on several measures. Which of the following would best fit the index's requirements?
   a) The number of electrical gadgets and other luxury items
   b) Life expectancy, literacy rate, and availability of clean water
   c) The number of cars per household
   d) Amount of waste generated

98. This communicable disease is said to cause the maximum number of deaths each year all over the world. It is often associated with malnutrition or the lack of quality food. Name the disease.
   a) AIDS
   b) Tuberculosis
   c) Measles
   d) Tetanus

99. Cantharis, also known as Spanish fly, is used today in herbal remedies to treat cystitis. It has also been used for treating baldness, scalds, and burns. Which well-known Greek healer used it to treat dropsy?
   a) Socrates
   b) Aristotle
   c) Plato
   d) Hippocrates

100. A 'Life Cycle Assessment' is the investigation and evaluation of the environmental impacts of a given product or service caused or necessitated by its existence. Such an assessment is also known as...
   a) Life cycle analysis
   b) Ecobalance
   c) Cradle-to-grave analysis
   d) All of the above

Answers

1. a) Bamboo
2. b) Anaerobic digestion
3. b) Endosulphan
4. a) Methane
5. a) Remote sensing
6. c) Both chlorine and fluoride
7. b) Carbon dioxide
8. a) Phosphates and polyphosphates
9. c) Natural selection
10. b) Pollen
11. a) Cell division
12. b) Aphids
13. b) 1973
14. b) Stones help crush food in the body
15. d) Natural selection
16. c) Bat
17. a) Cinnamon
18. b) Musk deer
19. b) Bacterial action
20. b) Thin and smooth: to protect against creeping plants
21. d) Elephant
22. b) Jurong Bird Park (Singapore)
23. b) Jim Corbett National Park
24. b) Oxpecker
25. b) Mutualism
26. a) Mangroves
27. c) Madras Crocodile Bank
28. d) Sea cow
29. b) Habitat Loss, Introduced Species, Pollution, Population, Over-consumption
30. b) Commensalism
31. b) Olive Ridley turtle
32. b) The lifting force is roughly perpendicular to the local flow field
33. b) Turning coal into syncrude

34. a) Thermal mass
35. a) Ranikhola
36. d) Breeder reactor
37. a) India
38. c) Italy
39. a) Compact fluorescent bulbs are more efficient
40. c) More heat
41. b) 60-watt bulb
42. c) Refrigerator
43. d) United States of America
44. b) Petroleum
45. a) Natural gas
46. a) Propane
47. d) Indian and Burmese
48. d) AD 79
49. d) Tsunami
50. c) A comet
51. d) All of the above
52. c) Petrol and electricity
53. a) Trains
54. b) SUVs (sport utility vehicles)
55. a) Pumice
56. b) Physical weathering
57. c) Fulgurite
58. a) The nuclear age
59. a) John James Audubon
60. a) GEO (Global Environment Outlook)
61. a) 2000 - Malmo Declaration
62. b) Decade of Education for Sustainable Development
63. c) 14–15 May
64. a) Wetlands
65. b) IFOAM
66. d) Brazil
67. a) Agricultural runoff
68. d) All of the above

69. d) All of the above
70. b) Dioxins
71. d) Asbestos
72. b) Matthew Maury
73. a) Asit K Biswas
74. c) Location on river valleys
75. b) Transpiration
76. b) Aluminium sulphate
77. b) Yangtze
78. a) Volga
79. b) Arsenic
80. b) By lowering the pH level
81. a) Sethusamudram
82. b) Chlorine
83. d) East
84. c) Deepwater Horizon oil spill/BP oil spill/Macondo blowout
85. b) Gulf of Mexico
86. a) Oil spill
87. b) Dioxins
88. b) Polychlorinated biphenyls
89. a) Natural pearl
90. c) N, N- Diethyl-3- Methyl benzamide (DEET)
91. a) Diisononyl phthalate
92. b) Meghalaya
93. b) Oxygen-releasing compound
94. b) IRS-P4
95. a) PET (polyethylene terephthalate)
96. b) 66½°N and 90°N
97. b) Life expectancy, literacy rate, and availability of clean water
98. b) Tuberculosis
99. d) Hippocrates
100. d) All of the above

Section II

Short Questions

Agriculture ❯❯❯❯

1. What is the name given to food produced without the use of conventional pesticides and artificial fertilizers?
2. Name the gigantic weed that reaches a height of 12 feet and has 35.5-inch-long leaves.
3. By what name is the inner part of a seed, where the embryo is formed, known?
4. What do we call a plant or other organism, which gets its food from dead or decaying organic matter?
5. The Great Irish Famine of 1847 was linked to which crop?
6. Under which field of science is the study of crop production and management of soil covered?
7. This wonder transgenic rice with high pro-vitamin A owes its unique property to a gene from a particular flower. Which flower are we talking about?
8. Which term best describes the fungi that grow on trees?
9. What do we call the layer of fatty substances covering the outer plant epidermis?
10. What are the spores produced by anthers of stamens known as?
11. Which plant was used by ancient Egyptians to make paper?
12. Commonly known as Aloe vera, this plant can be separated into two basic products. What are they?
13. Biomass can be gasified to produce a synthesis gas that is composed primarily of hydrogen and carbon monoxide. What is this gas called?
14. Soya bean or rapeseed oil, animal fat, waste vegetable oil or micro-algae oil produce which fuel?
15. Name this ancient spice used by the Chinese for ink and by the Romans as sandwich spread; while the Arabs thought it had magical powers.
16. Fuel, food, and fibre—the basic building blocks for an industrial society—all come from a single plant source. Can you name the plant source?
17. Potatoes contain two types of antioxidant carotenoids, which help the eyes, skin, and heart; and are particularly helpful for muscular degeneration. Name these two antioxidant carotenoids.

18. This toxin is a potent human—naturally occurring—carcinogen produced by *Aspergillus flavus*, a mould found in food products such as corn. It also acts as a potent liver carcinogen in rodents. Which toxin are we talking about?

19. By what name do we know the Indian crop that is sown in July–August and harvested in October?

20. What does the term 'olericulture' refer to?

Answers

1. Organic food
2. Giant hogweed (*Heracleum mantegazzianum*)
3. Kernel
4. Saprophyte
5. Potato
6. Agronomy
7. Daffodil (Narcissus)
8. Decomposer
9. Cuticle
10. Pollen
11. Papyrus
12. Gel (the leaf pulp) and latex (a bitter yellow exudate)
13. Syngas or biosyngas
14. Biodiesel
15. Sesame
16. Hemp
17. Lutein and Zeaxanthin
18. Aflatoxin
19. Kharif
20. Science of growing vegetables

Air »»

1. What is the full form of NAAQS?
2. Which type of fungus is propagated on fly ash deposits to make them arable?
3. As compared to other times of the day, morning air quality is generally better or worse?
4. What is the only major air pollutant not to show significant decline over the last 20 years?
5. The Fujita Scale is related to which weather phenomenon?
6. Two pollutants have to be present to form ozone. One is oxides of nitrogen. Which is the other?
7. What is the term used to describe air that is so dry that precipitation evaporates before it reaches the ground?
8. What is summer smog also known as?
9. Out of trees, fish, and hens, which one are most likely to benefit directly from increasing levels of carbon dioxide?
10. At what minimum wind speed is a hurricane categorized as category 5?
11. The burning of trees in tropical rainforests increases the amount of which gas in the atmosphere?
12. Which type of radiation gets filtered by the ozone layer?
13. What does dry ice become when it sublimates?
14. When a layer of warm air lies above a layer of cold air, it leads to an atmospheric condition known as…
15. First reported in China in March 2003, what is the full form of SARS?
16. Which gas is blown through a nuclear reactor to carry the heat away?
17. Which pollutant in air causes hay fever?
18. Mining of which mineral has resulted in a great deal of air and land pollution in Goa?
19. Which polluting gas is emitted by photocopiers?
20. Can you name this extremely toxic chemical that is released when chlorinated plastics like polyvinyl chlorides (PVC) are burnt?
21. In which season is the concentration of carbon dioxide in the atmosphere generally higher?

22. Which blue-tinged gas, a pollutant near the ground level, helps form smog and causes acid rain, but creates a lifesaving layer in the stratosphere?

23. The World Health Organization (WHO) says this gas causes up to 15 per cent of lung cancers worldwide. Name this gas.

24. What is the term to denote relative humidity exceeding 100 per cent?

25. Condensation on hygroscopic particles commences when the relative humidity is below 100 per cent. What is this phenomenon known as?

26. What name is given to the area of calm at the centre of a tropical storm?

27. If carbon monoxide enters the bloodstream this compound is formed. It inhibits the blood's ability to carry oxygen. Name the compound.

28. What do we call a snowstorm that is accompanied by low temperatures, strong winds, blowing, and drifting snow?

29. Majority of weather elements that affects the Earth occurs in one of the layers of the atmosphere where heating and cooling of air masses takes place. Which layer is this?

30. How many minutes does it take a hurricane to release more energy than all the world's nuclear weapons combined?

31. Which gas is the primary constituent of Mars's atmosphere?

32. Which planet of our solar system has the strongest magnetosphere?

33. Cyclones are formed in areas of low pressure or high pressure?

34. What name is given to a tropical cyclone of severe intensity that occurs in the Pacific Ocean?

35. Terrorists in Tokyo used the nerve gas Sarin to kill subway commuters. What is the other name for Sarin?

36. Leakage of which poisonous gas resulted in the Bhopal gas tragedy?

37. Pertussis is a very contagious disease of the mucous membranes that line the air passages. By which name do we know Pertussis better?

38. This hot, dry wind blows from the Sahara Desert and passes through Egypt, Lebanon, Israel, and Syria. Can you name this wind?

39. Chemicals that have adverse effects on human health or the environment are usually called HAP. What is the full form of HAP?

40. Two gases released from fossil fuel combustion combine with water vapour to cause acid rain. One of them is sulphur dioxide; which is the other?

41. Can you name the two major pollutants released from domestic fuel consumption?

42. Though super saturation is required for cloud droplets to sustain themselves, relative humidity rarely approaches 101 per cent, even in very wet clouds. Then, how do cloud droplets grow into raindrop size?

43. What are the two mechanisms responsible for producing precipitation from clouds?

44. Why does the moon sometimes look reddish or orange while rising or setting?

45. Point sources account for nearly 90 per cent of this 'criteria' air pollutant. Which pollutant are we talking about?

46. Exposure to this gas must be prevented—rather than treated—as detoxification is quite difficult due to the gas's insolubility. Which gas are we talking about?

47. Name this volatile organic compound (VOC) used in pesticides and cleaning products, emissions of which adversely affect both human health and the environment.

48. Combined with oxygen and moisture, this particular industrial emission settles on the surface of marble structures, corrodes the limestone, forming a fungus that experts refer to as 'marble cancer'. Name this pollutant.

49. Burning coal produces sulphur dioxide, an acidic gas that contributes to the formation of acid rain. There is a method to clean up the gas before it is released. What is the name of this method?

50. Can you name this acute lung disease, which is an occupational hazard, caused by exposure to organic dust, fungus, or mould?

Answers

1. National Ambient Air Quality Standard
2. Mycorrhiza
3. Better
4. NOx
5. Tornadoes
6. Hydrocarbons
7. Virga
8. Photochemical smog
9. Trees
10. 155 miles per hour
11. CO_2
12. Ultraviolet radiation
13. Carbon dioxide
14. Temperature inversion
15. Severe Acute Respiratory Syndrome
16. Carbon dioxide
17. Pollen
18. Iron Ore
19. Ozone
20. Dioxins
21. Winter
22. Ozone
23. Radon
24. Supersaturation
25. Solute effect
26. The eye
27. Carboxyhemoglobin
28. Blizzard
29. Troposphere
30. 10 minutes
31. Carbon dioxide
32. Earth
33. Low pressure
34. Typhoon
35. GB (O-Isopropyl methylphosphonofluoridate)

36. Methyl Isocyanide
37. Whooping cough
38. Khamsin
39. Hazardous Air Pollution
40. Nitrogen oxide
41. Carbon monoxide and suspended particulate matter
42. The hygroscopic nature of certain condensation nuclei helps cloud droplets grow into raindrop size
43. Collision-coalescence process and ice-crystal process
44. Dust in the atmosphere diffracts the light
45. Sulphur dioxide
46. Mustard gas
47. Glycol ethers
48. Sulphur dioxide
49. Flue gas desulphurization
50. Hypersensitivity pneumonitis

Biodiversity

1. What name is given to a plant that has adapted itself to grow in salty water?
2. Which type of mosquito is most often responsible for transmitting the yellow fever virus?
3. What are plants that grow high up on other plants, especially in tropical rainforests, called?
4. Unlike most fishes, sharks do not have these. What are we talking about?
5. Name this small bird that flies from the Arctic to the Antarctic and back again each year.
6. What is the term used to denote a group of jellyfish?
7. The soft texture and arrangement of feathers in owls lend their flight a special characteristic, making it easier for them to surprise their prey. Name this feature.
8. What are crepuscular animals?
9. These newly identified microorganisms thrive in unimaginable conditions, like boiling sulphur springs and polar ice fields. What are they called?
10. Which term is used to denote the phenomenon of emitting light by glowworms?
11. What do we call the state in which organisms maintain internal equilibrium?
12. This term is used for species at risk because of low or declining numbers or small range; but are not yet threatened species. Which term are we referring to?
13. Name this tiger reserve in Rajasthan located at the junction of the Aravallis and the Vindhyas.
14. This is the biggest fungus of all. A single living colonial growth of this fungus in the forests of Washington State, US, is believed to be between 500 and 1,000 years old. Name the fungus.
15. From which tree, found in rainforests, is quinine made?
16. *Loxodonta africana africana* is the largest of all land mammals. By what name do we know this animal better?
17. It is the only snake species found inside the Arctic Circle. It is also Britain's only venomous reptile. Name it.
18. In which National Park has Jane Goodall been studying chimpanzees for over three decades?
19. What is common to termite mounds and ultra-modern human habitations?
20. There are only two animals with trunks. One is the elephant. Which is the other?

21. These gymnosperms existed millions of years ago. Out of these, only one has survived. Its common name is the Maidenhair tree. What is its scientific name?
22. Which condition in animals is characterized by white skin and hair, and pink eyes?
23. Genes from which two animals were combined to create a creature called the geep?
24. Which fish is called a 'living fossil'?
25. Of the approximately 1.75 million species of life identified by scientists, about 20 per cent belong to the order Coleoptera. By what common name do we know the insects that belong to this order?
26. The peppered moth evolved its colour due to increased production of which pigment?
27. Which is the largest rodent in the world?
28. Name the world's smallest deer.
29. Which type of ecosystem acts as a seasonal home for migratory birds?
30. Which bird became extinct after existing as the most abundant bird species on Earth?
31. One of the more striking characteristics of the butterfly is its development from an egg to a beautiful adult. During its metamorphosis cycle, the resting stage is characterized by formation of a pupa. What is this pupa called?
32. A seahorse has an adaptation that helps it watch out for predators and prey without moving its body. Name this adaptation.
33. The protozoan parasite plasmodium causes a tropical disease that is usually transmitted to humans by mosquito bite. Which variety of mosquito are we referring to?
34. The horseshoe crab is related to which terrestrial insect?
35. It is the smallest bird of all. It is found in Cuba and the Isle of Pines. Name it.
36. What is the study of biological classification known as?
37. Which is the only termite-eating marsupial in the world?
38. Name the lone staple food the giant panda survives on.
39. Which endangered species is the only great ape of the Asian region?
40. What name is given to a plant that grows on another plant, rather than on the ground?
41. What term is used to denote a species that is introduced to a geographical area where it does not occur naturally?

42. Which type of organism synthesizes organic materials from inorganic raw materials?
43. Intensive cross-breeding among different species of dogs was conducted to create the ideal guard dog breed. What is this breed called?
44. What is the lower part, or underside, of a sea turtle's shell called?
45. Name this organ that is located strategically in front of the roof of a snake's mouth and acts as a chemical receptor.
46. Which creature is considered to have the most toxic venom in the entire animal kingdom?
47. What sort of a skeleton does an insect have?
48. What do we call plants or fungi that obtain their nourishment from dead plant or animal material?
49. Near which lake is the Nalabana Bird Sanctuary located?
50. What is the elastic, horny, middle part of the sole of a horse's hoof called?
51. These animals survive freezing temperatures by turning their body fluids into glycerol. Name them.
52. What are fruit-eating animals called?
53. The tropical regions of which continent have the world's richest diversity of palms?
54. The striking white colour of seagulls is produced by the reflection of light by irregularly distributed air-filled cavities. What is this phenomenon known as?
55. When retracted into their shells, snails cover the entrance with a 'trapdoor'-like structure. What is it called?
56. Which is the only sanctuary in the world where we can find the Asiatic lion?
57. What is the unique feature about the eardrums of tropical frogs?
58. What are mesoplankton?
59. What are tiny aquatic animals eaten by fish called?
60. Bud cap is a feature of this flower. The name of the tree to which this flower belongs means 'well-covered'. Name the flower.
61. Koalas and wombats are a little different from kangaroos. How do they differ?
62. Can you name this rather small, agile, weasel-like carnivore native to Africa, southern Europe, and Asia?

63. It is the world's biggest flower that grows in the jungles of Borneo and Sumatra in Indonesia. It is 1m in diameter and weighs up to 11kg. Name the flower.

64. What term is used to denote the taxonomic level in the Linnaean sequence that is above 'class' but below 'kingdom'?

65. The elongated digits of a bat over which its wings are spread represent a human feature very closely. Which feature are we referring to?

66. During fall, we find leaves of different colours, such as red, wine-red, and purple. What lends such colour?

67. There are only three types of venomous mammals. Two of these are male duckbilled platypus and several species of shrews. Which is the third?

68. Nepal is home to about 600 Great One-Horned Rhinos. In which National Park do most of them live?

69. Pollutants can be more concentrated in animals higher in the food chain because of...

70. Name this tundra animal, which is most threatened by climate change because of the annual migratory pattern it follows.

Answers

1. Halophyte
2. *Aedes aegypti*
3. Epiphytes
4. Bones
5. Arctic tern
6. Smack
7. Noiselessness
8. Animals that are active only at dusk and dawn
9. Extremeophiles
10. Bioluminescence
11. Homeostasis
12. Vulnerable species
13. Ranthambhore Tiger Reserve
14. *Armillaria ostoyae*
15. Cinchona
16. African Bush Elephant
17. Adder, or the common viper
18. Gombe National Park (Tanzania)
19. Air-conditioning
20. Tapir
21. *Ginkgo biloba*
22. Albinism
23. Sheep and goat (at Cambridge in 1982)
24. Coelacanth
25. Beetles
26. Melanin
27. Capybara
28. Leaf muntjac, or leaf deer
29. Wetlands
30. The Great Auk
31. Chrysalis
32. Highly mobile eyes
33. Anopheles
34. Spider
35. Hummingbird

36. Taxonomy
37. Numbat
38. Bamboo
39. Orangutan
40. Epiphyte
41. Exotic species
42. An autotroph
43. Dobermann
44. Plastron
45. Jacobson's Organ
46. Sea wasp (a box jellyfish)
47. External
48. Saprophytes
49. Chilka lake
50. Frog
51. Poikilotherms
52. Frugivores
53. Asia
54. Structural colour
55. Operculum
56. Gir
57. They are placed externally, behind the eyes
58. Plankton that grows in deep sea locations
59. Zooplankton
60. Eucalyptus flower
61. The kangaroo's pouch is on the front; while the koala's and wombat's pouches are on the back
62. Mongoose
63. Rafflesia
64. Phylum
65. The human hand
66. Anthocyanins
67. Solenodons
68. Royal Chitwan National Park
69. Bio-magnification
70. Caribou

Energy

1. Sharks can detect fish buried in the sand with the help of their special cells. What kind of energy can these cells detect?
2. What are small, solid, extremely energy-efficient light bulbs, popularly used for battery-powered items such as flashlights and headlamps, known as?
3. What are the sources of matter from which biomass energy is derived?
4. Coastal areas, mountain gaps, rounded hills, and open plains are best suited for the production of which type of energy?
5. What is current flowing steadily in one direction around a circuit known as?
6. What is the term that denotes energy losses in transformers and chokes due to an alternating current core?
7. Name the source of energy that 'drives' atmospheric motion.
8. Heat radiated from buildings located in an urban area is referred to as which effect?
9. Coke ovens emit a high concentration of which gas?
10. In 1931, the 100 kW Balaclava turbine was the first modern wind machine to provide power to an alternating-current utility grid. In which country was this large wind turbine located?
11. Which is the single largest source of commercial conventional energy?
12. In the US, the most common fuel used to power turbines and generate electricity is coal. True or false?
13. When a refrigerator is stacked with food, does it consume more electricity or less?
14. At the global level, which is the leading renewable energy source, used to generate electric power?
15. Which natural process is responsible for capturing the majority of energy made available to living things on Earth?
16. Which ancient civilization is thought to be the first to use hydropower?
17. What is the other name for 'black light'?
18. What is the most common non-commercial biological fuel used in majority of developing countries?
19. These power a space shuttle's electrical systems, producing a clean byproduct in the process. What are we referring to?

20. In a video camera, in what form does light energy get stored?
21. Name the pollutant that is the result of incomplete combustion of certain solid fuels such as bituminous coal.
22. Which fossil fuel generates the lowest amount of carbon dioxide?
23. What is the name given to rotors or blades that spin when driven by steam, gas, water, or wind?
24. Solar energy is increasingly becoming popular as an important source of renewable energy. Which country has the largest solar cooker programme in the world?
25. In what form do superconducting energy storage systems store energy?
26. Which type of energy storage system is usually built in abandoned mines?
27. Magnetic energy produced by gigantic storms on the Sun's surface strikes the upper atmosphere of the Earth, thereby producing colourful patterns in the sky near the Poles. What is this phenomenon known as?
28. In which country was the first large wind turbine connected to a modern utility?
29. Why is carbon dioxide gas blown through a nuclear reactor?
30. By what name do we refer to the process of exposing biomass to high temperatures in the absence of oxygen, causing it to devolatize?

Answers

1. Bioelectrical energy
2. LED (light emitting diodes)
3. Plants and animals
4. Wind energy
5. Direct current
6. Hysteresis loss
7. Sun
8. The heat island effect
9. Hydrogen sulphide
10. Ukraine
11. Oil
12. True
13. Less electricity (because the food soaks up the cold air and keeps it trapped inside)
14. Hydel
15. Photosynthesis
16. Greek civilization
17. Ultraviolet radiation
18. Fuelwood
19. Hydrogen fuel cells
20. Magnetic energy
21. Smoke
22. Natural gas
23. Turbine
24. India
25. Electromagnetic form
26. Compressed air
27. The Northern Lights or aurora borealis
28. Ukraine
29. To carry the heat away
30. Pyrolysis

Environmental Disasters

1. Which major volcano's eruption drastically affected solar radiation and brought about major weather changes all across the globe?

2. Which two countries in Asia were rocked by a powerful earthquake in October 2005?

3. This illness was the cause of maximum deaths in Europe during the medieval period. It spread to Europe through the trade route to Asia. Can you name it?

4. One of the most spectacular geological events is a volcanic eruption. Volcanoes form when molten rock rises to the surface from the Earth's core, crust, or the mantle?

5. Name the deadliest earthquake in recorded history, which is said to have taken more than 800,000 lives.

6. In 1960, in which continent did the worst earthquake of the twentieth century occur (measuring 9.5 on the Richter scale)?

7. Both reindeer in Lapland and sheep in Britain were affected as a result of the fallout from a nuclear disaster. Which incident are we referring to?

8. Which famous nuclear accident was the worst to have occurred in the USA?

9. The bubonic plague is a disease that spreads to humans via fleas from infected rodents. Name the germ that causes this plague.

10. On 15 April 2006 this river reached its highest level for approximately 111 years, forcing residents of Romania, Bulgaria, Hungary, and Serbia to flee their homes. Which river are we referring to?

Answers

1. Krakatoa
2. India and Pakistan
3. The bubonic plague
4. The mantle
5. The Shaanxi earthquake in China (on 23 January 1556)
6. South America (Chilean coast)
7. Chernobyl
8. The Three Mile Island accident
9. *Yersinia pestis*
10. River Danube

Fuel and Transport

1. When hydrogen is burnt as fuel, which byproduct is formed?

2. Ethanol is produced by fermentation of sugar. True or false?

3. Jackrabbit acceleration, which refers to the sudden acceleration of a motor vehicle from a standing stop position, uses a lot of fuel. It also causes harm to the environment. How?

4. In 2010, which company was the world's largest motor vehicle manufacturer?

5. In 2010, which city had the world's busiest Metro system, carrying an estimated 3.16 billion passengers every year?

6. When a diesel vehicle is converted to CNG, which is the only pollutant whose emission increases?

7. Converting the carbohydrate portion of biomass into sugar, and then fermenting it, produces which fuel?

8. Besides tailpipe exhaust, which is the other prominent source of pollution from vehicles?

9. What is transesterification?

10. Name this device, which is fitted to motor vehicles to eliminate some of the harmful gases produced by internal combustion engines. It reduces oxides of nitrogen to a mixture of nitrogen and carbon monoxide, while unburnt hydrocarbons are oxidized to produce harmless products.

11. This gas, used as a fuel to drive cars, not only offers better mileage than petrol but also produces the lowest emissions of toxic gases, low noise levels, while improving the life of the engine. Can you name this gas?

12. What happens to the byproduct that is produced while powering a space shuttle with hydrogen?

13. What happens when biofuel is added to petroleum products?

14. Lead is one of the two most toxic chemicals used in petrol that cause brain damage and leukaemia. Which is the other?

15. Which gas, found in vehicular emissions, is fatal for human beings at high concentrations, and was used during World War II by the Nazis for this very function?

Answers

1. Water
2. True
3. Vehicle produces higher emissions
4. Toyota
5. Tokyo
6. Methane
7. Ethanol
8. Evaporated gasoline
9. A process that turns vegetable oil and animal fat into esterified oil, which can be used in place of diesel fuel, or can be mixed with regular diesel fuel
10. Catalytic converter
11. CNG (compressed natural gas)
12. Used as drinking water for the crew
13. It allows the fuel to combust more completely, and thereby reduces air pollution
14. Benzene
15. Carbon monoxide

Land

1. Scientist have identified areas on Earth (such as forest cover and oceans), which have the capacity to absorb greenhouse gases and clean the air around us. What are these areas called?
2. Which is the most common element occurring in the Earth's crust?
3. What is soil with equal proportions of sand, silt, and clay along with organic matter known as?
4. This sedimentary rock was formed from accumulation of marine plants and animals. Name it.
5. What do we call a cluster of islands representing the tops of a submerged coastal mountain range?
6. Which is the highest mountain in the world in terms of total height (from base to peak)?
7. The world's largest caldera, or volcanic depression, is found in Tanzania. What is its name?
8. Thunderstorms and lightning deliver a large amount of which element to the air and soil?
9. Soil microbes use ammonium nitrogen in the soil for energy, and in doing so 'oxidize' ammonium nitrogen first into nitrite and then into nitrate, which plants use for growth. What is this process popularly known as?
10. Three major harmful chemicals are associated with gold mining. Two of these are sulphuric acid and cyanide. Which is the third?
11. Which is the largest active volcano in the world?
12. If you fall into a pool of quicksand what should you do: stay still or try to get out as quickly as possible?
13. Which desert contains the world's largest expanse of unbroken sand?
14. Which desert is considered to be the world's oldest?
15. Name the only desert where the Joshua tree grows.
16. Name the island that has been evolving from the remnants of the Krakatoa volcano in Indonesia.
17. There are seven major tectonic plates upon which the continents move. True or false?
18. Igneous rocks that form entirely beneath the Earth's surface are plutonic or volcanic?

19. The tectonic plates of Earth are part of the lithosphere or the mantle?

20. Which layer of Earth is solid, yet flows?

21. According to the Mohs scale for mineral hardness, which is the second hardest mineral after diamond?

22. What are metal-bearing rocks called?

23. Name this volcanic rock that is characteristically dark in colour, contains 54–61 per cent silica, and moderate amounts of iron and magnesium.

24. What do we call a column or pillar of bizarre shape caused by differential erosion on rocks of different hardness?

25. Name this volcanic rock that is characteristically light grey in colour.

26. What is the name of the famous active volcano on the island of Sicily?

27. Which element is depleted most from the soil after a crop is harvested?

28. What name was given to the area of central and southwestern United States during the prolonged drought of the 1930s?

29. What were the three basic rock types used in the construction of the Lincoln Memorial in Washington DC?

30. What do we call a change in population over a geographical area, corresponding to environmental change?

Answers

1. Natural sinks
2. Oxygen
3. Loam
4. Limestone
5. Archipelago
6. Mauna Kea (In Hawaii, it rises over 33,000 feet from the ocean floor)
7. Ngorongoro crater
8. Nitrogen
9. Nitrification
10. Mercury
11. Mauna Loa
12. Stay still (the grains of sand will settle down and make it easier for you to get out)
13. Arabian Desert
14. Namib Desert
15. Mojave Desert
16. Anak Krakatau
17. True
18. Plutonic
19. Lithosphere
20. Mantle
21. Corundum
22. Ores
23. Andesite
24. Hoodoo
25. Dacite
26. Etna
27. Potassium
28. The Dust Bowl
29. Igneous granite, metamorphic marble, sedimentary limestone (igneous granite was used in the base and lower steps of the memorial, the floor and statue of President Lincoln are made of metamorphic marble, and the interior columns and walls are composed of sedimentary limestone)
30. Cline

Personalities

1. Who was the first to patent incandescent and fluorescent light bulbs in England?
2. Name the British merchant who brought about a significant impact on food preservation with his 1810 patenting of the tin can.
3. Aristotle and Socrates wrote on skins of dead cows. Am I serious or joking?
4. Noted wildlife activist Bittu Sahgal is the founding editor of this environmental magazine. Can you name it?
5. Which English scientist discovered the existence of cells by studying a slice of cork under a microscope?
6. Which Swedish naturalist devised a system of classification based on patterns of similarity between different organisms?
7. Which scientist, who discovered that a protozoan called plasmodium causes malaria, was the first Indian-born Nobel Laureate?
8. Who is the father of modern toxicology?
9. The well-known book Earth in the Balance: Ecology and the Human Spirit, was authored by a Senator from Tennessee. Name this individual.
10. Henry David Thoreau is one of the pioneers of the environmental movement. What is the name of his most famous book?
11. Name the ship on which Charles Darwin travelled and later formulated the theory of evolutionary change.
12. Who propounded the Big Bang Theory?
13. Who is considered to be the Father of Modern Astronomy?
14. Who was the author of the story The First Men in the Moon?
15. What is Sir Fred Hoyle famous for?
16. Name the famous Finnish architect who designed the first skylights.
17. This man, through his cartoons, was the first to draw the attention of the masses to animals, thereby promoting a feeling of love and attachment towards them. Who are we referring to?
18. Which discovery by Madam Curie led to a breakthrough in the fight against the dreaded disease cancer?
19. Who proved the existence of large star systems or galaxies far beyond our own galaxy?
20. Who created the cartoon strip character 'Pogo' and coined the phrase 'We have met the enemy and he is us'?
21. In 1995, the Russian cosmonaut Valery Polyakov completed a record 438 consecutive days in space. What was the name of the space station where he lived?

22. Who is known as the Father of Medicine?
23. Who authored the classic Lifetide that dealt with life and its biological rhythm?
24. What was the name of the 10-year-old American girl who wrote to Yuri Andropov, the leader of the USSR, in the winter of 1982 asking him to stop the nuclear arms race?
25. Integrated Pest Management (IPM) has been the guiding principle and environmental conscience-keeper of global agriculture since the late 1960s, as a responsive sequel to the book Silent Spring. Who was the author of the book?
26. Which Austrian monk postulated a theory of heredity based on his observations of pea plants?
27. Who proposed the theory of natural selection?
28. Which famous Nobel Prize winning scientist was also an Olympic athlete?
29. Narmada Bachao Andolan (NBA) is a non-governmental organization that mobilized tribals, adivasis, farmers, environmentalists, and human right activists against the Sardar Sarovar Dam built across the Narmada river, Gujarat, India. Who initiated the NBA?
30. Name the famous personality who explored Africa from Lake Victoria down the River Congo to the sea.

Answers

1. Joseph Swan
2. Peter Durand
3. Serious
4. Sanctuary Asia
5. Robert Hooke
6. Carl Linnaeus
7. Ronald Ross
8. Paracelsus
9. Al Gore
10. Walden; or, Life in the Woods
11. HMS Beagle
12. George Gamow
13. Nicholas Copernicus
14. H G Wells
15. He developed the steady-state model of the Earth
16. Alvar Aalto
17. Walt Disney
18. Radium
19. Edwin Powell Hubble
20. Walt Kelly
21. Mir
22. Hippocrates
23. Lyall Watson
24. Samantha Smith
25. Rachel Carson
26. Gregor Mendel
27. Charles Darwin
28. Neil Bohr
29. Medha Patkar
30. H M Stanley

Policies

1. Which UN programme plays a major role in caring for the world's environment, and was constituted by the UN General Assembly in 1972 to promote environment science, information, and its management?
2. This convention lays down access to information and public participation in environmental matters. Name it.
3. In which year did the United Nations General Assembly adopt the world charter for nature?
4. How many Millennium Development Goals (MDGs) are there?
5. In 1968, the African Convention was drafted. What did it concern itself with?
6. This protocol of 1925 prohibits 'the use of asphyxiating, poisonous, or other gases, and of all analogous liquids, materials and devices' and also banned use of bacteriological weapons. Which protocol are we talking about?
7. A principle states that 'the costs of pollution be borne by the party responsible'. The practical implications of this principle lie in its allocation of economic obligations in environmentally damaging activities. Which principle are we talking about?
8. The Convention for the Conservation of Migratory Species of Wild Animals was formed in which year?
9. In 1987, the World Commission on Environment and Development coined the term 'sustainable development'. By what name is this commission also known as?
10. By what name is the International Treaty on Plant Genetic Resources for Food and Agriculture better known as?
11. The Convention on International Trade in Endangered Species of Wild Fauna and Flora (CITES) was drafted in which year?
12. When did India sign and ratify the Kyoto Protocol?
13. Which was the first country to ratify the Vienna Convention for the protection of the ozone layer?
14. The day of signing the Montreal Protocol was proclaimed by the United Nations General Assembly as The International Day for the Preservation of the Ozone Layer. On which date every year is this day celebrated?
15. In which city was the famous Earth Summit of 1992 held?
16. Out of the several principles of international environmental policy, some were first enunciated in the Stockholm Declaration. Later, these were crystallized through which declaration?
17. This 1992 convention provides a framework for the negotiation of detailed protocols on issues such as control on the emissions of greenhouse gases—particularly carbon dioxide—and deforestation. What is this convention called?

18. What is the International Convention for the Prevention of Pollution from Ships also known as?
19. The SOLAS Convention is regarded as the most important of all international treaties concerning the safety of merchant ships. Its first version was adopted in 1914. What is the full form of SOLAS?
20. Name the famous legislation enacted in 1972 by the government of India, which established schedules of protection of plant and animal species, and largely outlawed the hunting or harvesting of these species.

Answers

1. United Nations Environment Programme (UNEP)
2. The Aarhus Convention, 1998
3. 1982
4. Eight
5. Conservation of nature and natural resources
6. The Geneva Gas Protocol
7. Polluter Pays
8. 1979
9. The Brundtland Commission
10. The International Seed Treaty
11. 1973
12. August 2002
13. Canada
14. 16 September
15. Rio de Janeiro
16. The Rio Declaration
17. The United Nations Framework Convention on Climate Change (UNFCCC)
18. The MARPOL Convention
19. Safety Of Life At Sea
20. The Wildlife Protection Act of 1972

Waste

1. In 1885, where was the world's first waste-fired electricity generation plant opened?
2. The first-ever sanitation force was created in 200 BC in which city? It consisted of teams of two men walking along the streets to pick up garbage and then throw it in a wagon.
3. In 3000 BC, the first recorded landfill site came up, wherein waste was placed in large pits and covered with earth at various levels. Which city are we referring to?
4. In March 2002, which country released the National Waste Strategy, and became the first nation in the world to formally adopt a zero waste strategy?
5. Cathode ray tubes (CRTs) are picture tubes used in computer monitors and TVs. Which toxic substance is present in CRTs that can pollute the soil and groundwater if disposed of in landfills?
6. Name the canal in New York that was destroyed in 1953 by waste from chemical plants.
7. Which industrial sector generates the maximum quantity of oxygen-demand wastes?
8. What term is used to denote bacteria or fungi injected into compost to initiate a biological action?
9. What do we call the process by which solid clumps aggregate and are separated from water or sewage?
10. Which country is the global leader in garbage recycling?
11. Should a landfill be built above or below the groundwater table?
12. Which part of Australia accounts for the largest share of electricity generation from sugarcane waste?
13. Name this plant operation—a volume-reducing system—that burns garbage in an oxygen-free environment.
14. Burning of plastics emits a pollutant that is carcinogenic in nature. Name it.
15. On which date is Green Consumer Day celebrated?

Answers

1. Shoreditch, London
2. Rome
3. Knossos (capital of Crete)
4. New Zealand
5. Lead
6. Love canal
7. Food processing industries
8. Inoculum
9. Flocculation
10. Germany
11. Above
12. Queensland
13. Pyrolysis
14. Dioxins
15. 28 September

Water »»

1. What name is given to a regular tidal wave that travels up the course of a river from its estuary?

2. Which continent faced its first environmental disaster caused by an oil spill from an Argentine supply ship named Bahia Paraiso?

3. Rain is naturally acidic. What is its pH level: about 5 or about 9?

4. Which is the only east-flowing river of Europe?

5. What is the temperature (in Celsius) at which water has maximum density?

6. The Gulf Stream is an ocean current that flows along the Antarctic Ocean. Is it a warm or cold current?

7. What causes hurricanes and cyclones: sea temperature or lunar gravity?

8. This term, also denoting a Greek alphabet, refers to the part of the river before it meets a lake or the sea. Which term are we referring to?

9. When a river meanders, it often slits to form a shallow lake beside the river. What is this lake called?

10. Which is the largest non-point source of water pollution: agriculture or industry?

11. Which of the following tend to have larger absorption abilities: rivers or lakes?

12. What do we call the top of a water saturated zone?

13. By what name are reefs parallel to the shore but separated by a broad lagoon known as?

14. What do we call the deepest portions of ocean basins?

15. What is the process of deposition of soil particles at the bottom of lakes and streams known as? The soil forms a layer over fish food and fish eggs, and poses problems to their life cycle.

16. TMDLs are used to limit pollution in water. What is the full-form of TMDL?

17. Next to the seasons, what is the most powerful force controlling global weather?

18. A waterfall is formed when water meets soft rock or hard rock?

19. The Three Gorges Dam is proposed to be built on which Chinese river?

20. The marine ecosystem yields many useful products. Peanut butter and toothpaste are made from a compound found in the sugars of red algae. Name this compound.

21. Of the four major bodies of water on Earth, which is both the largest and the deepest?

22. Which toxic element is commonly found in old water pipes?

23. Which country pioneered the art of drip irrigation?

24. Jaundice is caused due to the pollution of this important resource. It is also responsible for more cases of human illness the world over than any other environmental factor. What are we referring to?

25. Which natural biological process uses substantial amounts of oxygen and carbonate, reduces water hardness, has a mild acidifying affect, and is beneficial for fish to thrive?

26. Name the term that denotes the relative clarity of water.

27. What do we call the habitat near rivers or streams for plants and animals?

28. What term, referring to or pertaining to the ocean, is applied especially to organisms that live at the surface of the ocean, away from the coast?

29. Where is most of the freshwater on Earth found?

30. What are small ice crystals formed in the centre of super-cooled liquid water droplets known as?

31. What are large, dangerous ocean waves produced by hurricanes or typhoons called?

32. Which disease is caused by the presence of excess nitrogen in drinking water?

33. Compared to the land surface, there is lot more water on our planet; yet there is scarcity of water. Why?

34. What is the name given to the portion of land that is drained by a river and its tributaries?

35. Coliform bacteria in water indicate the presence of ...

36. What is an underwater volcano called?

37. What is the primary effect of excess phosphorus in an aquatic environment known as?

38. What do we call a habitat described as an area where freshwater and salt water mix?

39. This exotic plant brought to India from another country causes eutrophication in water bodies. Can you name the plant?

40. Water harvesting has emerged as a sensible method of meeting water shortfall in a cost-effective manner. What does water harvesting mean?

41. Which is the highest waterfall in the world?

42. Where is water found on Mars?

43. Which ocean has witnessed the maximum number of tsunamis?

44. People suffer from amebiasis when they swallow water or food contaminated with stool from an infected person. This parasitic infection affects which part of the body?

45. Which invention made it possible for an individual to dive into the sea and ocean to explore and admire the underwater environment?

46. What do we use to measure the amount of precipitation that has fallen during a given period of time at a particular place?

47. With which ocean is the La Niña weather event associated?

48. The world's oldest lake, which is also the world's largest freshwater lake in terms of volume, holds one-fifth of the global freshwater. Can you name the lake?

49. How many rivers are there in Saudi Arabia?

50. This is a water-borne infectious disease prevalent worldwide. It is caused by parasitic flatworms found in tropical freshwater. To infect humans, the worms must first infect and mature in freshwater snails, which are their 'intermediate hosts'. Which disease are we talking about?

51. When the Suez Canal was opened, marine life migrated to the Red Sea and the Mediterranean Sea. What is this ecological phenomenon known as?

52. Located off the Pacific coast of Japan, this was once considered to be the deepest point in the world (subsequently found to be in the Mariana Trench). Name this point.

53. What is the scientific study of freshwater bodies for their biological, physical, and geological properties known as?

54. What is the perpetual movement of sand along a coastline called?

55. By what name do we know the 'land between the two rivers' Tigris and Euphrates better?

56. Name the largest delta in the world.

57. This is a manmade lagoon that was created by run-off from the Svartsengi power plant, which pumps up geothermally heated water from a full mile below the surface. After being used to generate both heat and electricity, the excess water (which is absolutely clean) is ejected into the lagoon. Name this lagoon.

58. The Red Sea got its name because of a type of algae. When these blooms of algae die, they appear to turn the blue-green colour of the ocean to a reddish-brown. Name the algae.

59. The Murray river is Australia's greatest river. Where does it originate from?

60. Although the River Saraswati does not have a physical existence today, there are numerous references to it in ancient Indian literature of the Vedic period. A part of the river exists in Haryana. What is it known as?

Answers

1. Bore
2. Antarctica
3. About 5
4. Danube
5. 4 °C
6. Warm
7. Sea temperature
8. Delta
9. Oxbow lake
10. Agriculture
11. Lakes
12. Water table
13. Barrier reefs
14. Ocean trenches
15. Siltation
16. Total Maximum Daily Load
17. El Nino
18. Hard rock
19. Yangtze Kiang
20. Carrageenan
21. Pacific Ocean
22. Lead
23. Israel
24. Water pollution
25. Nitrification
26. Turbidity
27. Riparian
28. Pelagic
29. Polar caps
30. Ice embryos
31. Storm surge
32. Methemoglobinemia
33. Only 1 per cent of the available water is drinkable
34. River basin
35. Human faeces

36. Seamount
37. Eutrophication
38. Beach
39. Water hyacinth
40. Collection of rainwater on rooftops and in storage tanks to recharge groundwater
41. Angel Falls (Venezuela)
42. North polar cap
43. Pacific Ocean
44. Large intestine
45. Aqualung
46. Rain gauge
47. Pacific Ocean
48. Lake Baikal (Siberia)
49. None
50. Schistosomiasis, or snail fever
51. Lessepsian migration
52. Tuscarora Deep
53. Limnology
54. Longshore drift
55. Mesopotamia
56. The Ganges-Brahmaputra-Meghna delta plain
57. Blue Lagoon
58. Trichodesmium erythraeum
59. The Snowy Mountains
60. Ghaggar

Miscellaneous

1. Which country in Europe is considered the greenest in terms of greenhouse gas emissions per capita?

2. A banana kept inside a refrigerator blackens prematurely. The cold air, to which the banana is exposed, causes the production of blackening compounds. What are these compounds known as?

3. Introduction of synthetic fibres in the 1950s replaced the most widely used material for making ropes. Which material are we referring to?

4. At the heart of the sun, every second, hydrogen converts into which gas?

5. Of the four states of matter, three are solid, liquid, and gas. Which is the fourth?

6. Can you name the light that is million times brighter than sunshine?

7. What is the main disadvantage of using high-pressure sodium lamps?

8. Which term is used to denote converting coal into syncrude?

9. Which type of plastic is used for making wires with plastic coating?

10. PET is used extensively as a packaging material. What is the full form of PET?

11. The synthesis of which vitamin in human skin is triggered by UV rays from the Sun?

12. Which disease is transmitted by *Aedes aegypti*?

13. Which common term is derived from the Greek word 'plassein'?

14. Name nature's most abundant protein polymer.

15. The Indian television series Living on the Edge was the first in Asia to win the prestigious Panda Award. By what popular name do we know this award?

16. There are three major types of mercury found in the environment. Methyl mercury, mercury zero, and mercury two. Out of these, which one is the most toxic?

17. This disease among humans is caused by parasitic worms called Schistosomes. Can you name the disease?

18. Which is the principal component of the rigid cell wall that most plant cells have, in addition to the plasma membrane?

19. Which chemical is used in deodorants as the sweat-blocking ingredient; and hence, is unsafe?

20. The first phonographs, or LP records used a substance obtained from which plant?

21. What is the term used to denote an interacting group of individuals belonging to the same species?
22. Which term is used to refer to the morphological, physiological, anatomical, and biochemical characteristics of an individual?
23. What do we call the unit of chromosome that codes for specific proteins?
24. Which term is used to denote different forms of the same gene?
25. Which is the densest planet of our solar system?
26. Which metal can be recycled over and over again without any loss in quality?
27. This chemical used for bleaching paper, produces dioxins when in use. The recycled paper is thus not environment friendly. Which chemical are we talking about?
28. Name the rare colourful clouds formed about 65–70 km above the Earth's surface that can only be seen at night.
29. Also known as 'K', what is the term for the maximum maintainable population of an area?
30. Which African fly carries the trypanosome parasite that causes sleeping sickness in humans?
31. Radon is a gas derived from the natural radioactive decay of which metal?
32. What term corresponds to 'summer', if 'hibernation' corresponds to 'winter'?
33. What do we call the science of study of annual rings of a tree to ascertain its age?
34. Rusting of materials containing iron on exposure to water and oxygen is a process of...
35. What term is used to denote computer software that is used to map the Earth?
36. Which colour would you associate with a defoliant, which was used commonly by the US forces in the Vietnam conflict?
37. Which element is said to be the reason behind the Mad Hatter's madness in Alice in Wonderland?
38. What name is given to the science that deals with the universe beyond the Earth?
39. Which metal must necessarily be present in an amalgam?
40. Which is the world's highest active volcano?
41. Name the world's first cloned mammal.

42. Which is the simplest type of hydrocarbon?

43. What is the scientific study of ants known as?

44. Why does the Eiffel Tower grow taller by 15cm every summer?

45. Which country is the largest producer of a small, oval fruit that has a thin brownish-green skin and a fuzzy surface? The name of the fruit is directly linked with the country.

46. Due to a ban on DDT, there was a rise in this disease in certain third world countries. Which disease are we talking about?

47. When the moon changes from new to full, it gets bigger. What is the term used to denote this process?

48. What is the name given to a man who does scientific mapping of the moon?

49. What are sounds below 20 Hertz referred to?

50. Of the following, which is the most unpredictable: hurricane, earthquake, or volcano?

51. Which pathogen is associated with amoebic dysentery?

52. This is a bacterial infection in humans. Transmission is associated with contact with infected animals or water contaminated with rat urine. It is also known as Weil's disease. Can you name it?

53. Name this theory that states the universe originated from a gigantic explosion of a central large and dense mass.

54. Which term is used for a group of individuals who promotes the research, education, and conservation of primates?

55. Which is the first country to insert an amendment into its Constitution, thus allowing the state to protect and improve the environment for safeguarding public health, forests, and wildlife?

56. Astronomers estimate the accurate age of a planet by using rock samples from the same planet. Can you name this scientific method?

57. Which theory states that new matter is continually being created at various parts of the universe?

58. How many calories are equal to one pound of fat?

59. What is the meaning of the Latin word 'nebula'?

60. What do we call the full moon occurring just before the autumnal equinox on 23 September in the north temperate zones?

61. Which element is believed to be responsible for the decline of the Roman Empire because it affected the mental capacity of Romans?

62. What is the scientific study of ice, snow, hail and other low-temperature phenomena known as?

63. This is one of the three principal animal deities of Hindu mythology that has evolved after the Vedic Period in Indian history. Which country has its national airlines named after this deity?

64. The cut-flower trade began in the Netherlands in the 1600s. Name the flower that started this trade.

65. What is the study of the conditions necessary to support the independent development of extraterrestrial life called?

66. Nuts are a rich source of copper and magnesium, and are an excellent source of a particular vitamin, which is known to be beneficial to the heart. Name this vitamin.

67. Alphaviruses, flaviviruses, bunyaviruses, and reoviruses are important classification of which type of virus?

68. In March 1998, a genetic engineering technology was patented in the US to sterilize seeds of certain food grains. It turned out to be the biggest and most controversial news in seed research. What is this technology popularly known as?

69. During a battle between the Incas and the Spaniards in 1534, the volcano Cotopaxi erupted and put an end to the fighting as both parties fled from the battlefield. Where is Cotopaxi located?

70. Psittacosis, also known as ornithosis, is a disease humans can contract from birds infected with *Chlamydophila psittaci*, a bacterial-type organism. What is the common name for this disease?

71. This is a potentially fatal disease that is transmitted by the bite of an infected tick. A common infection in animals, it is caused by the Babesia parasite. Which disease are we talking about?

72. What do we call seeds that do not survive drying and freezing?

73. Which planet spins so fast that it bulges at its equator?

74. Which subject deals with the possible environmental influences on the development of species, especially that of man?

75. Some illnesses transmitted by insects do not produce symptoms until long after the infecting bite. Onchocerciasis is one such illness that can result in loss of sight years after the initial infection. What is this disease commonly known as?

76. Name the galaxy in which our solar system is situated.

77. What is the study of wildlife sounds called?

78. Which is the closest galaxy to Earth that can only be seen from the southern hemisphere?

79. This chemical is used in the manufacture of certain high-quality, clear plastics (for example, feeding bottles). It is also used in the resin that lines most tin cans. Which chemical are we talking about?

80. Exposure to prolonged intense solar radiation (usually in highly reflective environments) causes inflammation of the cornea. What is the scientific term for this painful eye condition?

81. Name this foodborne disease that is caused by a toxin produced by the bacteria *Clostridium botulinum*. It affects the nerves; and if untreated, can cause paralysis and respiratory failure.

82. What is the broadened bottom of a tree trunk, which helps balance a tree, called?

83. Which asteroid is of exceptional interest to astronomers because it comes closer to the Earth than any other body of comparable size, apart from the moon?

84. Organophosphates stop a key enzyme in the nervous system from working, making people ill. Name this enzyme.

85. Why do fireproof clothing often have a shiny surface?

86. This disease is caused by the largest and only protozoan with cilia or hair-like structures. It infects the large intestine usually after consumption of food or water contaminated with human or animal faeces. Name this disease.

87. Which is the world's most common plaster, which is, in essence, composed of processed straw?

88. What is the collective term used for herbicides, avicides, algaecides, rodenticides, fumigants, and so on?

89. What is the other name for pepper spray used by law enforcement agencies to control mobs?

90. This is a seismically active region of the Earth located in the Pacific Ocean. What is it popularly known as?

91. In ancient times, this disease was mentioned in Egyptian hieroglyphics, and was described in detail by the Greek physician Hippocrates. Which disease are we referring to?

92. Gamma-Hexachlorocyclohexane (HCH) is a nerve poison and an organo-chlorine pesticide. But it is often used for the treatment of head lice and scabies. What is its common name?

93. Onions emit an irritating substance when we peel them that makes our eyes water. What is this substance?

94. Visible to the naked eye even under moderate conditions, this galaxy was called the 'Little cloud' by the Persian astronomer Abd-al-Rahman Al Sufi. Name the galaxy.

95. Historically, Indonesia and Japan have accounted for the maximum number of active volcanoes. Which country ranks third?

96. Sun-damaged skin, or sunspots, medically called solar lentigo, is caused by exposure to the Sun. What is the other name for sunspots?

97. Why does stainless steel not corrode?

98. What is the technical name for bio-control, a system that uses natural means, rather than chemical pesticides, to control pest population?

99. What is the name given to the area around the Taj Mahal, over which a blanket ban has been imposed regarding establishing industries using fossil fuels?

100. The moon and the Earth are approximately at the same distance from the Sun. Then, why do they not have the same temperature?

Answers

1. Switzerland
2. Polyphenals
3. Manila hemp
4. Helium
5. Plasma
6. Laser
7. They contain mercury, a toxic substance
8. Coal liquefaction
9. PVC
10. Polyethylene terephthalate
11. Vitamin D
12. Dengue
13. Plastic
14. Collagen
15. The Green Oscar
16. Methyl Mercury
17. Bilharzia
18. Cellulose
19. An aluminum compound
20. Carnauba palms
21. Population
22. Phenotype
23. Gene
24. Allele
25. Earth
26. Aluminium
27. Chlorine
28. Noctilucent clouds

29. Carrying capacity
30. TseTse
31. Uranium
32. Aestivation
33. Dendrochronology
34. Oxidation
35. GIS (Geographic Information System)
36. Orange
37. Mercury
38. Astronomy
39. Iron
40. Nevado Ojos del Salado (Chile)
41. Dolly, the sheep
42. Methane
43. Myrmecology
44. Due to thermal expansion
45. New Zealand, hence the name kiwi fruit
46. Malaria
47. Waxing
48. Selenographer
49. Infrasonic
50. Earthquake
51. Protozoa
52. Leptospirosis
53. The Big Bang Theory
54. Primatologists
55. India
56. Radioisotope dating method
57. Steady State Theory

58. 4,000 calories
59. Cloud
60. Harvest moon
61. Lead
62. Cryology
63. Indonesia (Garuda Indonesia)
64. Tulip
65. Methane hydrate
66. Vitamin E
67. Arbovirus
68. Terminator Technology
69. Ecuador
70. Parrot fever
71. Babesiosis
72. Recalcitrant seeds
73. Earth
74. Paleobioclimatology
75. River blindness
76. Andromeda
77. Bioacoustics
78. The Clouds of Magellan
79. Bisphenol A (BPA)
80. Photokeratitis
81. Botulism
82. Buttress
83. Eros
84. Cholinesterase
85. To reflect the radiated heat away from the body
86. Balantidiasis

87. Cow dung
88. Biocides
89. Oleoresin capsicum
90. The Ring of Fire
91. Malaria
92. Lindane
93. Oil containing sulphur
94. Andromeda
95. USA
96. Liver spots
97. It contains small amounts of nickel and chromium
98. Integrated Pest Management (IPM)
99. The Taj Trapezium Zone
100. Due to the moon's lack of atmosphere

About the Educating Youth for Sustainable Development (EYSD) Division, TERI

The involvement of today's youth in implementation of environment and development programmes is critical to the long-term success of Agenda 21. Enhancing the role of youth and actively involving them in the protection of the environment and the promotion of economic and social development are the basis for positive action. The Educating Youth for Sustainable Development (EYSD) Division at TERI has initiated several projects and activities with the aim of providing today's youth with a secure and healthy future. The overall focus of these programmes is to provide a 'quality' environment with improved standards of living. Maximum engagement of youth is sought through thematic projects that empower the youth so that they can influence and catalyse environment improvement through effective resource-management initiatives—both at the local and global levels.

EYSD firmly believes that there is a need for governments, civil society organizations, academic institutions, and every individual of this planet to come together in solidarity for the protection of the only planet where life is possible. In order to realize this belief, EYSD works with the following stakeholders, besides the youth:

- School teachers and wider schools community comprising principals, administrators, grade-IV employees and parents
- National and international educationists
- Policy makers and leaders
- Various national NGOs, INGOs, UN bodies, and funding agencies
- Multinational corporates with a strong commitment to sustainable development
- Education boards, institutions, and universities
- Marginalized and underprivileged communities
- Educational and cultural networks, youth groups, and trainers

The Division is also involved in cutting-edge research on Education for Sustainable Development, and brings out books, newsletters, workbooks, manuals, films, and other IEC materials periodically.

For more information, log on to http://www.teriin.org/themes/youth/youth-theme.php